SPEAKING OF CATS

By Harry Cauley

J. N. Townsend Publishing
Exeter, New Hampshire
2003

Copyright © 2003 Harry Cauley. All rights reserved. No part of this book may be duplicated without written permission from the publisher.

First Printing.

Published by
J. N. Townsend Publishing
4 Franklin Street
Exeter, New Hampshire 03833
800/333-9883.

www.jntownsendpublishing.com

ISBN: 1-880158-45-0

Library of Congress Cataloging-in-Publication Data

Cauley, Harry.
 Speaking of cats / Harry Cauley.
    p. cm.
SIBN 1-880158-45-0
  1. Cats--New York--Germantown--Anecdotes. 2. Cats--Behavior--New York--Germantown--Anecdotes. 3. Cauley, Harry. I. Title.

SF445.5.C414 2003
636.8'0887--dc21   2002041626

A special thank you to Sally-Jane Heit for giving the manuscript to Julia Lord and to Julia for sending it to Jeremy Townsend who has made this an especially rewarding publishing experience. And also to Terri Fyler, the title miester, for "Speaking of Cats."

For my sisters, Ann and Mary, with lots of love.

A cat may look on a king. – John Heywood

FOREWORD

For some unknown reason, there is no mention of a cat in the Bible. I don't know about the Koran or the Talmud, but the Bible doesn't have a single solitary reference to a cat. Dogs, yes, and cattle and sheep and goats and horses and on and on and on. It even mentions the pygarg. What's a pygarg, you may ask, aside from a very good word to remember when you're playing Scrabble? Biblical scholars think it was probably some kind of antelope. Whatever it was it is mentioned in the

Bible. But no cat. Lions, of course, are mentioned over and over but your plain old everyday cat didn't make it. If that's not enough to make a cat insecure, I don't know what is.

They have been domesticated for about 5,000 years as opposed to dogs, who moved into the caves with early man about 50,000 years ago. The cat, apparently, needed another 45,000 years to decide whether or not to trust us. Since then, life for the cat has certainly had its good and bad times. In ancient Egyptian and Norse religions they were venerated. In the Middle Ages they were tortured and burned as witches. Even Shakespeare couldn't seem to make up his mind about cats. "A harmless necessary cat," although bland, seems pleasant enough but then, in *All's Well That Ends Well*, Bertram says, "... I could endure anything but a cat, and now he's a cat to me ... A pox upon him! For me, he's more and more a cat." So, if they appear schizophrenic off and on who can blame them.

But times change and even if the Bible couldn't work a cat into its text, Andrew Lloyd Webber's *Cats* is the longest running show in the history of Broadway. So, there!

The diametrically opposed attitude

Speaking of Cats

toward cats will go on until the end of time. On the one hand the pro-cat people indiscriminately say, "I just love everything about them." But on the other, the cat detractors think it's OK to say any negative thing they want. It seems no one worries about being politically correct about a cat.

"I just hate them."

"They're always sneaking around. They give me the willies."

"You never know what they're going to do."

"I'm allergic to them."

"I hate the way their tails move. You know what I mean? Like snakes."

"They think they're better than people."

"They've got the devil in them. Everybody knows that."

It's no wonder it took them 45,000 years longer than the dog to accept us.

Having said that about the two camps I have to admit that I am not exclusively a cat person but I am an animal freak. My mother always said I liked anything that could walk, wiggle, swim or fly. I have in the past shared my life with goats, an orphaned sheep named Lollipop, ponies, turkeys, chickens, ducks, geese, guinea fowl, peacocks, a skunk named Gus and a crow named Ernie as well as innumerable dogs and cats. They were all creatures of such distinct and diverse personalities that I have never been able to relate to them with any degree of scientific detachment. A victim and product of Uncle Remus, Beatrix Potter, Walt Disney, Pogo and *The Wind in the Willows*, I am a devout anthropomorphist, certain when no one's around that rats and mice sing "Cinderella" and cows stand on their hind legs and dance like the Rockettes while they moo in harmony. So, the reader is warned. The cats I write about were my friends and I think about them as being as wise, foolish, loving and complicated as most of my two-legged friends. I hope I have captured some of what and who they were.

> "All right," said the (Cheshire) Cat; and this time it vanished quite slowly, beginning with the end of the tail, and ending with the grin, which remained some time after the rest of it had gone.
> –Lewis Carroll

SOCRATES

I grew up in a catless world. Well, at least a catless neighborhood. Almost everyone had a dog but I have no recollection of anyone having had a cat. And there certainly was no place for a cat in my family because my mother had had a very bad experience with a cat when she was a little girl. An obnoxious cousin threw an old tom on her back when she wasn't looking. The frightened animal probably dug his claws in to hold on, so she was terrified of cats for the rest of her life. She wasn't really an animal person to begin with but she was tolerant. She had no choice, because my dad raised English setters and pointers and I always had my own house dog. In the backyard there were pigeons, rabbits and all kinds of wild abandoned babies.

The first cat came into my life once I was out of the army, on my own and working. I was living in Philadelphia, in Germantown, in a crumbling old brownstone that was only a few years away from the wrecking ball. There were six very large apartments in the building but it was in such disrepair that only two were occupied, illegally I'm sure, mine on the third floor and the one right below me.

It was in the early sixties, before we were all too paranoid to leave our windows open for fear of being robbed or worse. Since my apartment was on the third floor and there was no porch roof, no tree, no fire escape, no possible way anyone could get in the window, I always left it open in good weather. Even if someone went to the trouble of throwing a grappling hook up and pulling himself through the window, which would be difficult to do without attracting attention, there was nothing to steal. My furnishings were as sparse as a monk's cell. I did have a bed and a nightstand with a good reading light but there was no dresser. I kept my clothes in my beat-up old army footlocker. I had a kitchen table and two chairs in the spacious dining room and a mattress with pillows that served as a sofa in the living room. It was draped with the requisite Indian-print cotton throw. And there was a fairly decent coffee table, which I had found on the street. Travel posters, unframed of course, were taped to the walls. It was an amalgam of *La Bohème* and beatnik but it more than served my purposes and I thought it was great.

One early autumn evening I came home from work at a bookstore and decided to take a quick nap before going to a movie. I stretched out on the bed and just as I started to doze off my unfocused eyes wandered around the room. There, sitting on the windowsill, was a cat. I thought I was seeing things. There was no possible way, short of flying, he could have gotten to the window. I bolted upright, startling him almost as much as he startled me. I was dumbfounded. He was a light-gray, longhair with just the shadow of tortoise markings, as handsome a cat as one is ever likely to have suddenly appear on his windowsill. His legs seemed disproportionately long and thin, like a teenager who had reached his full growth but had yet to fill out. He stared at me with unblinking dark-green eyes, almost as though I was the curiosity intruding on his territory. It was a standoff, we were both waiting for the other to make the first move.

"Where'd you come from?" He was so captivating I expected an answer. If he could fly surely he could speak. At the sound of my voice he blinked, which I interpreted as the beginning of negotiations. I stood up and immediately put him on the alert. When I moved toward him his ears went back and he opened his mouth in a soundless hiss. It was clearly a warning. I respected it and stepped back and sat on the bed again. That was the distance he was comfortable with so his ears went up and he seemed to settle a bit.

"Now what do we do?" I don't know how long we sat watching each other before he relaxed enough to take a quick swipe at cleaning himself. His front paw went up to his mouth and with the toes curled under he started licking. I knew nothing about cats but was certain that if he was comfortable enough to give himself a bath it was a good sign and I decided right then that if he wanted to stay it was all right with me.

Friends of mine who lived in apartments had cats so I knew the drill. Litter pan, litter, food dish, water dish, dry food, and canned food. If he was going to stay he wouldn't make it through the night without needing litter, so I went shopping and when I got back to the apartment he was still sitting on the windowsill.

"Like it or not, you're staying. I have over twenty bucks invested in you. That's a commitment." He yawned.

No way was I going to have a litter pan in my bedroom so I decided to put everything he needed in the doorway to the hall. Once he started using the facilities, I would eventually move the litter pan to the bathroom and the food and water to the kitchen. When everything was in place I went to the bed and sat down to wait for his first foray into my world. I started talking to him. I told him about my job at the bookstore, trying to impress him by making certain he knew I was the manager. I went on to tell him how I had pretty much given up my aspirations to be an actor, about my family, my mother's cat phobia, where I grew up, the animals I had when I was a kid, how much I was enjoying finally reading *War and Peace*, anything I could think of. After listening to me for what seemed like hours and deciding the only real threat I posed was boring him to death, he jumped off the sill into the bedroom and tentatively started investigating. It was the first time I got a really good look at him. He was huge. Huge and truly beautiful. One of the biggest cats I had ever seen and I started to have second thoughts about having him around the apartment because I knew that if he ever decided to turn on me I would lose the battle. The investigation went on without so much as a glance in my direction, probably because he was confident he had the upper hand. And, if worse came to worse, he could just fly out the window on his invisible cat wings.

I have never seen so much nose and whisker twitching. Since there was no furniture except the bed, nightstand and footlocker he concentrated on the footlocker, which had been up and down the east coast and in Germany and I'm sure had the best stories to tell. Finally, when he sniffed as much as he felt was necessary and computed all the information, he sat about eight feet away and resumed his staring. I had the spooky feeling he was trying to decide exactly what to do with me. I started talking again.

"OK, so *War and Peace* is about these Russians..." and this time he let me know what he thought about my babble by going straight to the litter pan and using it. I perceived this as an editorial comment so I shut up. He knew what a litter pan was so he wasn't quite the wild and savage thing I suspected he might be and that was reassuring.

For the next few months we went through an intensive period of adjustment. And, like most people with cats in their lives, I adjusted to everything he demanded. He was a no-nonsense cat with no apparent sense of humor, arrogant and wary and not the best roommate I ever had. At times I seriously thought about getting rid of him but I didn't know how. I suspected he thought a lot about getting rid of me, too.

It was the first time I was introduced to the fussy-eater syndrome. Dogs, for the most part will gratefully, if indiscriminately, eat anything in their dish. Cats sit in scrupulous judgment of everything you put before them as though they were food critics for some gourmet cat magazine. There were certain foods he ate and certain ones he tried to cover up as though he had just used the litter. Some food he sniffed endlessly before cautiously taking a bite, testing it for poison, and others that made him bypass the kitchen completely for an entire day. My life became a quest for the perfect food for this itinerant, uninvited and, for a good deal of the time, unwanted guest in my home.

And he was not to be touched. Just as royalty is not to be touched. When I put my hand out, although he never scratched or bit me, he made it perfectly clear that I might lose an arm if I tried to pet him. He instantly assumed the Halloween cat posture. The ears went flat, he arched his back and the emerald eyes narrowed as he opened his mouth and snarled the soundless hiss. I retreated to safer ground and licked my wounded ego, accepting the fact that he was never going to be my best friend and wondering just why the hell I was keeping him.

However limited our relationship may have been, he was a fascinating creature to watch. Every movement echoed his bigger cat cousins. Just walking down the hall or jumping up on the windowsill he became the lion, tiger and leopard, fearsome and disdainful yet elegant and graceful beyond measure. And, while

Speaking of Cats

I watched him I devoted hours to wondering where he came from and what had happened to him in his short life. Why was he so cautious? Had he been abused? And the biggest question of them all was how did he get to the window? Even if he could have talked he probably wouldn't have told me. The Sphinx would have been more talkative.

I think he must have slept all day while I was at work because except for dozing on the windowsill, and even then he seemed to keep one eye open, I never saw him sleep. Whatever time I finally went to bed, he went to a spot he had chosen in the bedroom doorway. I put a throw rug down for him, but rather than sleep on it he moved a few feet to the left of it. He was definitely a fellow who marched to his own drummer. There was a little night-light in the hall which cast a soft glow on him so I could watch him while he gave himself his bath every night and I never stopped marveling at how fastidious he was. There is nothing as mesmerizing as watching a cat bathe. All that repetitive stretching, bending, curling and licking until he had each hair in place exactly the way he wanted it was downright soporific. I don't think I ever managed to stay awake long enough to see him actually finish.

He had, among others, one obviously peculiar characteristic. He never made a sound. Never in all the time we were together did he make a single meow. For all intents and purposes he was mute. The closest thing to sound was the rush of air through his throat when he hissed and that was only when he felt his territory was being invaded. I don't know if the silence was the result of something organic or by choice.

How and when I gave him a name, I don't remember. I have no recollection of trying to come up with one that suited him, no discussion with friends over a bottle of wine, no watching for some little idiosyncratic aspect of his personality that might suggest a name. One day he was Socrates and that's all there was to it. It seems a bit pretentious in retrospect but I was young. I had been to Europe, I managed a bookstore, I read Kerouac and Sartre and dressed mostly in black. I had poetry readings in my apartment, I went to the theater and listened to classical music. If I didn't have a right to be pretentious, then who did?

When Socrates and I had been together about six months I had weekend guests. Two of my nieces, Kathleen, eight at the time and her little sister, Colleen, who was six, came for a visit. After months of being nagged because I had casually invited the girls to come to Philadelphia, my sister and brother-in-law took them to the station, and, I'm sure with great fear and trepidation, put them on the train to ride alone for one hour. It was a much safer world in those days. All alone with no grownups to tell them what to do, it was bliss and they have both been fiercely independent ever since. They were beaming when they got off the train in Philly. It was back when people dressed for travel, and the two of them with their hats and white gloves and new little suitcases were the hit of the platform.

Before their arrival I had two considerations. One, how would Socrates react to them and two, how could I give my sparse and rather depressing apartment some degree of charm so it would appeal to two little girls? As far as Socrates was concerned, I'd have to stay on my guard and wait and see. As far as the apartment was concerned, Marguerite, a woman who worked with me in the bookshop, had the solution. I went to a store that sold Asian paraphernalia, bought dozens of paper lanterns, a few strings of Christmas lights and went home and made magic. One of the benefits of living in a building that's about to be torn down is the freedom to pound nails in the walls anywhere you want. By the time I finished, the place looked like it was ready for the Chinese New Year. I put the girls in the bedroom, which was transformed by the soft glow of the lanterns into something akin to an exotic brothel, and I slept on the sofa/mattress in the living room, which looked like a third-rate sushi bar. I had one chrysanthemum in a cheap vase on the table, a nice touch I thought, which gave new meaning to the word "minimalism," and I told the girls my decor was Asian. The Asian culture was utilitarian and therefore only used what was absolutely necessary. They fell for it hook, line and sinker and they loved it. I knew they would go home and ask their parents to redecorate. We ate Chinese take-out with chopsticks sitting around the coffee table, walked along the river, went to the zoo and decided we really didn't like zoos, went to two movies and decided we really liked movies, ate pizza and sang a lot of songs. I taught them how to tango. It was a perfect weekend.

As far as Socrates was concerned I told

them he was temperamental and it was best to stay clear of him. That was fine with Colleen, the younger one, who inherited her grandmother's reluctance with animals, but Kathleen was like me and I could see she was itching to get her hands on him. And he was fascinated with them, although he never came out where they could get a good look at him. He lurked in the shadows and around corners, as he always did when there were strangers in the apartment, and kept his eye on them every moment. But I never felt there was anything threatening about him. He was merely curious.

The second night they were there, done in by all the activity, they were ready to go to bed at eight o'clock. So was I. I tucked them in and went back to the sushi bar and stretched out with a book and fell asleep before I turned the first page. It's much easier to work a full day in a coal mine than it is to find ways to entertain two kids. Even kids who were easily pleased and great fun to be with.

It was about ten when I woke up and thought I had better check on them, so I tiptoed down the hall and saw that Socrates wasn't in his usual spot on the floor. I panicked and immediately thought of all the bizarre cat horror stories where the breath is sucked out of children during the night. All rational thought leaves when you're responsible for other people's children and, in a split second, your mind becomes a creep show. I don't know what I expected to find when I pushed the door open but my heart was pounding and I was breaking into a sweat. A blood-curdling scream was just waiting in the wings. And there he was on the foot of the bed, not asleep but comfortably curled up, looking at me as though I were an intruder. The girls had wanted the lanterns left on, so they were all bathed in a warm sepia glow and the three of them looked perfectly beautiful. It was Norman Rockwell at his best. I didn't take it well at all. I was jealous. Never once, in six months, had that cat gone anywhere near the bed when I was in it. I felt leprous and untouchable. As far as I was concerned he was a user and a traitor and right then and there I started thinking about getting rid of him. To make matters worse, he made me feel petty and small because I was jealous of my little nieces. I didn't need another conscience, I had enough trouble with the one I already had.

The next day I took the girls to Mass and

brunch and then to the train station. I was as reluctant to see them leave as they were to go back to the mundane reality of their young lives; school, nuns, home, baby brother, father and mother authority and everything else that life conspires to inflict on a kid. How would they survive it all after two days in Shangri-La? Home and hearth could only pale by comparison. After tearful good-byes and enough waving from the windows for a farewell scene in a World War II movie, the train pulled out and I headed for my Datsun.

Driving back to Germantown I thought about Socrates and tried to put everything in perspective as objectively as I could. He was nice to look at but he didn't really like me and I was starting not to like him. I had to admit he wasn't much trouble and it was pleasant having company in the apartment. Someone to talk to. But was it worth it? He was just a cat. It's not like we discussed the latest books or politics or religion. I fed him and cleaned his litter and in return he stared at me and slept with strangers. I didn't need that kind of rejection. It was possible for him to live another fifteen years and the prospect of the two of us barely tolerating each other for fifteen years was not very inviting. By the time I got home I had decided that if I could catch him, I'd try to find a home for him. I would start all over again. I would get to know cats by going to the animal shelter and adopting a kitten. Well, that was my plan.

When I got to the apartment I avoided making any eye contact with him because quite early on in our relationship I decided that along with sensing my moods, he could read my mind. And, if he knew I was thinking of finding him another home there was no telling what he might do to me or how he might dispose of my body.

I put off taking down the lanterns. Living in a sushi bar wasn't half-bad. The girls left everything in perfect order, even making their beds, which I'm sure was a last minute instruction from my sister. I had nothing to do for the rest of that Sunday but avoid contact with Socrates. It wasn't easy. He knew something was up and he was watching my every move. The staring was intense and I could only take so much of it, so I called a friend and we met for dinner and a movie. When I got home that night I opened the door and he was sitting on the coffee table waiting for me. It was something he had never done before and it was very scary. Peculiar behavior in an animal is unnerving and I had read that when wild animals act odd, it could be a sign of rabies. Why not a slightly domesticated cat? So, by the time I went to bed I was almost convinced that there was a rabid cat in the house whose plan was to do away with me.

Once I was settled in bed, Socrates went to his usual spot in the doorway and I relaxed a bit. But, just to keep me off balance, he didn't take his bath. He sat staring at me until, worn out from the weekend, I could fight sleep no longer and I drifted away. I don't know how much later it was when I felt the slight movement at the foot of the bed. I looked down and he was there, starting his bath. It was a moment of triumph and I wanted a brass band to play as I went to the window and shouted out to the world that my cat was on my bed and life was good. I wanted to thank him for this extravagant show of affection, but instead I contented myself with watching his ritual bath, knowing we were finally, to some degree, friends.

And so we started a more relaxed, if not exactly intimate chapter in our life together. For the next several months we lived in peaceful coexistence. I still couldn't touch him but it didn't matter. Whatever degree of trust I inspired in him was such a great gift and so flattering, I didn't expect anything more. He waited for me at the door every night and all but rubbed up against my ankle when I came home. That would have been asking too much. We were happy together; undemanding, supportive and respectful of one another. Man and cat had come to terms and the treaty was signed. It was an oddly satisfying and, on many levels, very solid relationship.

Someone once said we are all just a phone call away from catastrophe or euphoria. I don't want to think about that too much because it makes day-to-day living very edgy and it's tough enough as it is. But I know a single phone call can change the direction of your life forever. That's certainly happened to me many times. I had long ago decided to forget a career in the theater and be as happy a bookstore manager as I could. My life was fine. I liked living in Philadelphia; I had good friends, intellectual stimulation, enough money to get by, a car that was usually in running condition and a cat that slept on the foot of my bed. And then, the phone rang. All it took was one fifteen-minute conversation with an old friend and her new husband, Marte and Bill Slout, who had a summer theater in Michigan, asking if I was interested in being their leading man, and my life was changed forever. It was so long bookstore, hello fulfillment, elation, camaraderie, insecurity, petty jealousies, great parts, lousy parts and unemployment lines. I am eternally grateful to them for coming to my rescue and putting my life back on track. It's amazing what one phone call can do.

After giving my notice and milking my friends for several farewell lunches, dinners and a great party, I got down to the business of preparing to leave. I still had a few weeks so I did all the things that had to be done for a major life change. I closed my meager bank account, got my hair cut, had the Datsun tuned up, bought new underwear and a cat-carrying case for Socrates. When I brought it home he eyed it suspiciously, looked at me for a moment and then disappeared under the bed. I had never seen him run from anything. What did the carrying case mean to him? It was a clue to his past but I knew it was still a mystery that would never be solved.

When I came home from work on my final

day, which was more emotionally wrenching than I had anticipated, Socrates wasn't waiting for me at the door. I called him, but he didn't respond and when I searched the apartment I couldn't find him anywhere. He wasn't in the kitchen waiting to be fed, he wasn't sitting on the windowsill watching the comings and goings in the street below and he wasn't under the bed. He was nowhere in the apartment. He was gone and I felt sick. I went downstairs to my neighbor and asked if she had seen him but she hadn't. I went out and walked the streets calling him until almost ten o'clock at night and finally went home sure that he'd be waiting for me at the door when I got there. He wasn't. Again I searched the apartment but there was no Socrates. When I finally went to bed I coud not fall asleep. The apartment seemed too big and too empty and the soft glow of the lanterns, instead of making it warm and cozy, gave it an eerie, other-worldly feeling. When I finally did go to sleep it was fitful and I kept waking up expecting to see him at the foot of the bed giving himself a bath. All night I told myself he'd be back. And why not? He simply appeared before so why couldn't he disappear and appear again? Everything's possible when you're curled up in the dark trying to get through the night. But in the cold light of day I had to face up to it. He was gone.

The next morning I went through the neighborhood again and posted signs offering a reward and giving my phone number. A week went by and no one called. I had to leave or I wouldn't make it to Michigan in time for the first rehearsal. I had been so preoccupied with Socrates that I hadn't made any arrangements for getting rid of the bed and whatever else was in the apartment, so I just walked out and left everything. I packed my belongings in the car, every minute watching for a gray streak out of the corner of my eye. I was still thinking that somehow he would know I had to leave and at the last minute show up.

Before I closed the apartment door behind me I turned on all the lights, filled his bowl with food and put down fresh water just in case he came back. I stopped at my neighbor's but she wasn't home, so I slipped a note under her door with a phone number where I could be reached if and when he showed up. Then I went out front, got in the car and turned the key. The motor started but I couldn't pull away from the curb. I had the feeling he was watching me. Somewhere under a bush or a parked car or from under a porch those intense green eyes were watching me, wondering why I was leaving him. I wanted him to know that I was not abandoning him. I had to go. I had no choice. Finally, I put the car in gear and with the empty carrying-case on the back seat, headed for the Schuykill Expressway. I never saw him again.

As quickly and mysteriously as he had come into my life, he left. When I let my imagination run wild I think of him almost as a phantom cat. A ghost cat that might have lived in the apartment when the building was new and full of life. Maybe he was just visiting because he had some unfinished business. Maybe that's why he never let me touch him. It seems as good an explanation as any.

The funny thing is, when I ask my nieces about the weekend they visited me they remember so much. The lanterns, the Chinese food, the tango ... but they have no recollection whatsoever of a big gray cat.

When I play with my cat, who knows if I am not a pastime to her more than she is to me?
—Michel Etquem Montaigne

MARLENE KAZAMCHAK

Whether or not I believe in fate I really don't know. I add that to an increasingly long list of things I don't know. And the older I get the longer the list gets. Fate, Karma, Kismet, Destiny ... call it what you will. It seems to me that certain things are meant to be, certain people are meant to meet and certain events are meant to happen, because the odds against those people meeting and the events happening are so enormous the only explanation is, well ... fate. I suppose, when it suits my purposes, I want to believe that some things are preordained. As I say, I really don't know, but one of the greatest cats I ever lived with came into my life under such incredible circumstances that fate seems to have had a hand in it.

It was the end of the first season of our summer theater in Western Pennsylvania. My two partners and I had opened the doors in early June and expected to have every seat in the house sold. We promptly discovered that it took years to build an audience. It was the hardest work I've ever done, but the most rewarding. In the thirteen years that we had the place, we did some truly wonderful shows and I met people that are still an important part of my life.

At the end of that first season, upon closing up the theater for the winter, I headed back to New York, where I was living at the time. I was exhausted and not only broke, but heavily in debt. I barely had enough for gas, tunnel tolls and a few meals. Driving along Route 22, well over the speed limit, ignoring the beauty of the autumn foliage because I was preoccupied with how we were ever going to pay the bills we had incurred over the summer, I saw a kitten sitting by the side of the road. It was in the middle of a long, lonely stretch of highway where there wasn't a house for at least ten miles either way. I pulled over, backed up and got out to look along the berm until I found it. What made seeing the kitten so extraordinary was that she was so incredibly small. Somehow, miraculously, in the trash and weeds along the highway, going 75 miles an hour, I saw this kitten who was little more than half the length of a hot dog bun. What else but fate? She was barely alive, mewing pathetically and blindly at anything that moved. She had had to crawl up a hill to get to the side of the road but somehow she made it. I suppose it was the noise of the traffic that attracted her. When I picked her up she started sniffing my hand and bumping against my palm looking for a nipple. She was so tiny she almost looked embryonic. At first I wasn't even sure it was a kitten, but I know wild critters don't give birth in autumn. If she was old enough for her eyes to have opened, and I don't think she was, they were mattered shut and oozing. She was sitting in the palm of my hand in the warm sun, yet was shivering uncontrollably. This was the saddest and most forlorn little creature I had ever seen. I got back in the car and headed down the road. What was she doing all alone in such an isolated spot? It didn't occur to me until months later that had I looked down the hill behind her I probably would have found a bag or box with her brothers and sisters. If we are meant to be custodians of all the animals, as the Bible tells us, more often than not we do a really lousy job.

The next town was little more than a crossroad but there was a veterinarian, a crusty old guy who primarily tended farm animals, who offered to put her to sleep. There was no way I was going to let anyone put this kitten to sleep, humane as the vet's suggestion may have been. I was going to do everything I could to help her make it. No question about it, she was my responsibility. Fate. So, he cleaned her eyes and gave me some milk and a syringe without the needle to feed her until I could get a bottle. There wasn't anything else he could do except to tell me not to get my hopes up. I offered to pay but he very kindly told me to forget it and gave me some solution to bathe her eyes every so often. He probably thought I was as pathetic as the kitten. We tried to feed her some of the milk but she ate very little before falling asleep in my hand. The vet just shook his head as he showed us out of his office. I often wondered how many times over the years he told the story of the sap with the doomed kitten and I regretted not having his name and address so I could have written and told him the outcome of our little drama.

There was another cat in my life at the time, a seal point Siamese named Nutsy who was so gentle, kind and loving and so brave that he merits a chapter of his own which will come later. But I have to mention him here because he was instrumental in the survival of the tiny sick kitten. When I got back to New York I fed her what little she ate with a doll's bottle. But tender loving care was as necessary for her survival as food and what I had to give her wasn't enough, so Nutsy provided the rest. He would lie perfectly still while she sucked on a little tuft of his fur, as though she was nursing, until she fell asleep. He cleaned her and looked after her, as any good mother cat would do. So much for the stories about all tomcats killing kittens. Animals are as peculiar and individual as people with one great difference... they're considerably more dependable.

For the first week I was certain she wasn't going to live. She ate barely enough to survive. But I knew she was a fighter and I wouldn't give up hope. Her eyes cleared up but I could not get her interested in eating. A vet in New York gave me a few different formulas to try but one didn't seem to interest her any more than the other. She seemed to be more satisfied nursing Nutsy's tuft of fur. Then, one morning, for no apparent reason, she was famished. She couldn't seem to get enough to eat. Had this happened on one of those popular medical TV shows a doctor would have come out to the waiting room and said, "She's turned the corner. Everything is going to be all right." She certainly had turned the corner. She was still wobbly on her spindly little legs, but she was as playful as any kitten her age. Although she didn't seem to grow, she was equally as demanding when she was hungry.

Just as a gosling imprints on the first thing it sees when it hatches, thinking that is its mother, she imprinted on me. I was her private property the minute I picked her up on the side of the road and, for whatever little effort I made to help her survive, the reward was enormous. Nothing beats that kind of unconditional love.

She was a little calico cat, white with tabby gray markings and a small spot of tan under her chin and on her back right leg. Even though she eventually got over her eye problems, I later discovered that her pupils didn't dilate at all. Not that it seemed to affect her in any way, aside from bright light causing her to squint. She was as adept at catching bugs and mice as any other cat. But the huge, dark eyes with the yellow rims staring out of that little round face gave her a chronically inquisitive and startled look.

If ever an animal had reason to be wary and untrusting, considering the circumstances of her beginnings, it was she. But, aside from being shy with strangers once in awhile, she was happy and well adjusted. I give Nutsy most of the credit for that. His parenting skills were immeasurable.

Because she was abnormally small I always called her Little Cat and that was the name that stuck. She was Little Cat to one and all for the rest of her long life. However, she did have an alias. When she was about four years old, she was sitting on my lap at a barbecue in the apple orchard behind the summer theater when one of my co-producers, Nancy Chesney, who was also the leading lady of the company, tied a paper napkin around Little Cat's head like a babushka. She looked like one of those old photos of Eastern European immigrants ... like one of those pretty, round-faced, wide-eyed girls, full of wonder and ready to take on the New World. Well, being a writer I had to name this new character, this Eastern European cat immigrant that had suddenly appeared in the orchard. I don't remember if Nancy came up with the name or I did but I'm humbly going to take full credit for it. The name Marlene Kazamchak was perfect. So, Little Cat was occasionally Marlene K. or Ms. Kazamchak. She answered with equanimity to both names even though her bug-eyed stare always made her look like she was hearing them for the first time.

My life over the next several years certainly had its ups and downs and Little Cat and the rest of my animal family went along for the ride. We lived in a fifth-floor walkup—among several other cold-water flats—in New York City, the summer theater, a place on the beach in Virginia where she could run through the sea oats on the dunes, a charming and picturesque log cabin in Pennsylvania that had mice aplenty, a house in the Hollywood Hills with a pool and the place I now live in California. There never seemed to be a problem adjusting to any of our new digs. Home was wherever I unpacked the bags and, since cats are naturally curious, every new place was a great adventure. Aside from the moves to new houses, she even went on the road with me when I toured in a show and was a great little trouper as well as a spoiled favorite of the company.

She also had a theatrical career of her own. She was a shooting star, however, and burned out quickly. We were doing a play, which called for a cat and Little Cat, being as docile as she was, seemed the perfect candidate for the part. She didn't even have to audition. Believe me, it pays to sleep with the producer. There was only one scene in which she was involved, a cameo role you might say, and all she had to do was sit on Nancy's lap. That was it, nothing too taxing. Since she liked Nancy there didn't seem to be anything to worry about. Opening night came and she was superb. She captured every nuance and subtlety the role demanded. Hers was the work of an avowed method actor. She even took a bow with her leading lady in the curtain call. The second night she was equally as good, maintaining the high level of performance she had established. It was the Wednesday matinee that did her in, ending her bright but brief career. We had a full house, with the dependable ladies who came to all the matinees smelling deliciously of talc and perfume, so that when the curtain opened the fragrance wafted up on stage. The day was sweltering, over ninety degrees and the humidity at one-hundred percent plus, so we left the barn doors open.

I must say the matinee ladies were great audiences. Not only were they appreciative, but they were good sports as well. I remember one late day in September, a Wednesday, when it turned very cool, uncomfortably cool, and the ladies had come to the playhouse unprepared for the sudden change in the weather. Since the barn wasn't heated the best we could offer them were coats and capes from the costume department which they were happy to get. We had a small but grateful audience of ladies who looked like dress extras in a bizarre historical movie.

But the day Little Cat's theatrical career ended it was hotter than Hades. She had only been on stage at night so the audience was nothing more than a big black void that was oh-ing and ah-ing at how cute she was. But in the afternoon, with the barn doors open she could see all those people, all of those ladies fanning themselves with their programs. She wanted no part of it. The scene had barely started when she decided to quit show biz. She jumped off Nancy's lap, off the stage and into the audience, then ran out the door to freedom. Nancy was left with egg on her face and cat hair on her skirt. Luckily, the play was a comedy. The ladies were delighted and Little Cat, like Greta Garbo, who suddenly turned her back on her career and ran away from her public, became a legend in her own time.

I often thought Little Cat's mission in life was public relations for cats, because even the staunchest cat-hater fell victim to her charms. My mother, who never got over her fear of cats, actually sat on the sofa with Little Cat in her lap and petted her without breaking into a sweat. I was living in Los Angeles at the time and my parents, who were visiting from New Jersey, stayed on eastern time all the while they were there. They were usually up and showered and sitting having their second cup of coffee by the time I got out of bed every day.

The morning Little Cat sat in my mother's lap my father woke me.

"Skip, get out here. You've got to see this.

You're not going to believe your eyes."

We hadn't had one of our usual California disasters; no earthquake, no fire, no flood and the helicopters weren't circling the house, so I couldn't imagine what it was. I threw on a robe and raced into the living room trying to clear my head of sleep so I could handle whatever calamity had befallen us and there sat my mom with Little Cat in her lap. It was freaky. I will say that she didn't look very comfortable, sitting there straight-backed petting the cat with one rigid finger, so I knew Little Cat had not made an actual convert to orthodox catism. But it was a start. Sort of like watching an Arab and a Jew or an Irish Protestant and Catholic having lunch together. She definitely had an impact on my mother. After they went home, whenever I talked to her on the phone, which was almost every Saturday morning, between telling me what a delicious pot roast they had the night before and asking me what I was having for dinner, she always asked, "How's my friend, Little Cat?"

It wasn't long after my parents left to go back to New Jersey that I had a dream that my father and I went on a camping trip across the country in a jeep. I think we were going west to prospect for gold or maybe we were archeologists. Sort of Indiana Jones and his father. And the car was loaded with camping equipment, shovels, axes and everything else an archeological dig might call for. Also we had every dog and cat I had ever lived with in the jeep. Think, "Noah's jeep" and you'll get the picture. Even the first dog I could remember as a kid, Queenie, a bossy little terrier, who was already getting on in years when I was born, now mercifully a mere pup, was sitting atop all the equipment yapping her way across the country as any dog should in a fantasy adventure. The animals, like everything else in true dream fashion, were wonderfully skewed, larger than they were in real life, animated and very talkative. And very musical. We were singing our way through mesas and buttes right out of the best travelogues on TV when we started seeing the ubiquitous signs for a trading post. *Shorty's Trading Post. Moccasins, turquoise jewelry, live snakes, fossils, cold drinks, cactus candy and authentic Indian blankets.* They have been a blight on the landscape since the 1920s, but as much a part of the modern west as phony cowboys and Indians. The dogs and cats, especially a noisy Little Cat, were complaining about the signs and pointing out that animals do not deface the earth. As I recall, she was like any rabid and over zealous activist, pretty obnoxious about it. The next thing we knew we were in Shorty's Trading Post and Shorty, scruffy and wiry as any sidekick in an old western, was showing us around. He offered us some rattlesnake meat, which we both declined, and asked if we wanted to sell the dogs and cats. He told us he sold the freshest dog and cat meat west of the Mississippi and showed us some in the freezer section of the trading post (don't forget this was a dream), all packaged neatly like hamburger in a supermarket. The next part, the really good part, I don't remember in precise sequence, but somehow dad and I got the shovel and the ax and started trashing Shorty's. We were super avengers, defenders of all animalkind, capable of any feats of strength. I remember picking up a barrel, a barrel that would require a crane to lift off the floor, and throwing it through the plate-glass window. I was sensational. Indiana Jones would have been proud of us except, maybe, for our clothes because by then we were in full cowboy dress; boots, chaps, white hats, the works. Apparently some costumer had come in between sequences and dressed us appropriately for the big musical comedy finale. After Dad threw Shorty through the window, with all the animals cheering, we burned the place to the ground, got back in the jeep and headed west, into a perfect sunset, singing, "Don't Fence Me In," at the top of our lungs. It was one of my favorite dreams and I remember it as clearly as the one in which I won the Academy Award. Or was it the Tony? I'm not sure which, but I do recall my acceptance speech, which was brilliant.

Aside from her eyes with the pupils that didn't dilate, there was another result of Little Cat's shaky beginning. I don't mean to imply she was freakishly small or in any way stunted. She wasn't. She was quite nicely proportioned for a small cat. But she didn't have an estrous cycle and she never came into heat. I was very good about neutering my animals, unless, of course, I wanted to breed them. I admit I had trouble doing it to older dogs I adopted; more then likely it was a matter of transference on my part. But, for some reason I never had Little Cat spayed when she was still a kitten and then, when there was no sign of her coming into heat, I didn't think it was necessary. Over the years she certainly met enough randy tomcats in her travels that would have happily impregnated her if she had been receptive, but it never happened. It was a non-issue.

When she was about seven years old I noticed she seemed to be getting fat. I attributed it to middle-age spread. Again, it was a matter of transference. I was heading for my middle years and my waist certainly wasn't what it was when I was twenty. Since pregnancy was not even a possibility with Little Cat, I didn't pay any attention to her weight gain. One night I was lying in bed studying lines for a play and absentmindedly petting her when I felt something move in her stomach. Gas? It was the only explanation. But it moved again and again as though whatever was in there was trying to get comfortable. It took me awhile to accept the fact that, as impossible as it may have been, Little Cat was pregnant. I took her to the vet the next day and he confirmed it. He wasn't sure but he thought there were probably two kittens and I could expect their arrival in about a month.

Like everyone who is into animals I looked forward to the birth of newborns with delight. The very words for baby creatures; kitten, puppy, chick, duckling, foal, calf, kid, piglet, cub are enough to warm the heart. For most people, that is. I do have a friend who likes cats and dogs but his affinity for the animal world ends there. He was visiting one spring day when I had just had a clutch of mallard ducklings hatch and, like any proud parent, I was showing them off. I don't think I'd get much of an argument if I said newly hatched ducklings are about as endearing as an animal can get.

"Ugly little things, aren't they?"

My first reaction was to assume he was joking but I could tell from the look on his face he wasn't. He really thought they were ugly. How in the name of Hans Christian Andersen could anyone think a duckling was ugly? The man was clearly deranged. Hadn't Walt Disney built a preposterous fortune with a mouse and a duck as his cornerstones? And my friend thought the newly hatched ducklings were ugly! I automatically went through the stages of grief, because as far as I was concerned he was dead. Denial: No one could possibly think a duckling was ugly. Rage: The idiot should he shot, and so on until I got to Acceptance: Our friendship was definitely over, but charity prevailed and I knew I had to forgive his ignorance. Not, however, before I had my say.

"You are an insensitive lout who wouldn't recognize enchantment and charm if it came up and bit you on the butt." That's the printable version. What I actually said was peppered with much saltier language. Believe it or not, we're still very close.

Little Cat's pregnancy went on considerably longer than the vet thought it would. The month came and went and it was almost the end of the sixth week when she started looking for a place to nest. I fixed a box with a soft towel in it and placed it near my bed but it didn't merit any more interest than a quick sniff in passing. She looked in closets, behind doors and even in an old duffel bag, but nothing seemed to strike her fancy and she always returned to my bed. I had pretty much accepted the fact that my bed would be her birthing pad when, in the middle of the night, I was awakened by the sound of her arranging the towel in the box to her liking and I knew she had gone into labor. I turned on the light and prepared to help in any way I could although I knew that, generally, animals handle giving birth quite well without the interference of anxious and impatient people. I made myself a cup of coffee and waited and waited. Early in the morning, just as it got light and started to drizzle, she delivered one little kitten that I was sorry to see was stillborn. I waited, hoping the vet was right when he said he thought there were at least two, but she was only carrying the little dead baby. She licked it for awhile, trying to give it life, but realized it was hopeless. She jumped out of the box and onto my bed and started cleaning herself before settling down for a much-needed sleep. Living was about life. For her, it was the natural order of things. She moved on.

When I was certain there was no hope I buried the kitten in the apple orchard. All I could think of was the scene in *Tess of the D'Urbervilles* when Tess, in the attic with her younger siblings, baptizes her dying illegitimate child and names him Sorrow. I thought Sorrow was an appropriate name for the kitten because I still had the image of the pathetic Little Cat, sitting at the side of the road mewing for someone, anyone, to help her. Here was what would be her only offspring, born dead and being buried unceremoniously in the apple orchard with me, Nutsy and one of the dogs in attendance. There was something painfully desolate about it for me, but not for Little Cat.

I think animals, with the exception of human beings who are burdened with the terrible awareness of their own mortality, don't look back and don't wonder about the future. You can't say to a dog that wants to go out, "I'll take you for a walk the day after tomorrow," or to a cat that's hungry, "Remember the nice bits of steak I brought from the restaurant the day before yesterday?" For them there is no tomorrow and there is no yesterday. It's all about "now." But we, on the other hand, relive the good and the bad over and over and over for the rest of our lives. We contemplate the future with sometimes numbing fear or unrealistic optimism so much that we don't have time to enjoy the moment. We don't know the "now" nearly as well as animals do.

Little Cat lived happily for another ten years and I'm sure the dead kitten never crossed her mind once although it's still a vivid, sad memory for me.

Which one of us had the advantage?

I met a man with seven wives,
Each wife had seven sacks,
Each sack had seven cats,
Each cat had seven kits:
Kits, cats, sacks and wives,
How many were there going to
St. Ives?
—Anonymous

MRS. FISKE AND ROSEMARY

Actually, I have only purchased one cat in my life and that was the aforementioned and yet to be eulogized Nutsy. Aside from him all the cats I've lived with were foundlings or gifts or cats who, through no fault of their own, suddenly needed a new home. One of them, a beautiful blue point Siamese, was given by a patron of the summer theater whose children were allergic to anything with hair, fur or feathers. These were great kids who were crazy about animals so they had to be content with fish, reptiles and amphibians, on which they lavished as much love and attention as they would have on any warm-blooded pet. I don't know how the Siamese came to them but they passed her along. I certainly didn't need another cat in my life at the time because I was already sharing my digs with Nutsy and Little Cat. Not to mention a dog or two. But when they brought the soft gray kitten, as pale as a wisp of smoke, which looked at me with wondrous blue eyes filled with apprehension, I was a goner.

She was a shy little thing ... for all of about twenty minutes. Never have I seen an animal adjust so quickly to a new environment and take over so shamelessly. Nutsy was his usual tolerant self but Little Cat was less than pleased. This intruding nuisance was clearly not to be trusted. For one thing, she couldn't seem to do anything right. When she used the litter she was so energetic and enthusiastic about her digging that grit was tossed all over the bathroom floor and it looked as if there was more outside the box than in. When it was time to sleep she wanted to play. When it was time to eat, even though they all had their own bowls, she pushed the already disgruntled Little Cat away and ate whatever was in her bowl. It was the old story of two women living under the same roof. There was war. The new cat, in true Siamese kitten fashion, was blithely unaware of any wrongdoing and got into anything and everything she could. What is it about the Siamese gene that makes them not only vocally obtrusive but also absolutely fearless? Those qualities combined with their natural intelligence makes them creatures to be seriously reckoned with. And, they seem to show off all the time. At least she did. She always wanted to be the center of attention, so when it came to naming her the obvious choice was a theatrical one. A friend, the very funny author and actress, Fannie Flagg, had a Siamese named Tallulah, after the legendary and notoriously uninhibited Tallulah Bankhead, so that, the first name that came to mind, was already taken. At the time, half the Siamese cats living with actors were named Pyewacket after the cat in the play, *Bell, Book and Candle*, so it would have been like naming the kitten Jane Doe. So, I asked the members of the company, actors, technicians, everyone involved, to suggest a name. There was any number of considerations that made it to the finals. Bette Davis ... too much attitude, Ginger Rogers ... I already had a pair of geese named Fred and Ginger; Sarah Bernhardt ... a little too heavy for a kitten to carry; and Ethel Merman, who almost won the competition because of the kitten's vocal prowess. As I recall, Lucille Ball was also in the running. But nothing seemed tailored to this particular kitten. And then, I read an article in a theater magazine about an actress who had been the toast of Broadway in the late nineteenth and early twentieth centuries, Minnie Maddern Fiske, beloved as a comedienne as well as interpreter of the intellectual drama of her day. What more could anyone ask? It seemed to have the right ring to it. A little whimsical, a little classy, endearing and easy to shorten to a name the cat would remember. Minnie Maddern Fiske. And so, she became Minnie on her good days and Mrs. Fiske when she got into trouble.

As I've said, Nutsy was tolerant, much more interested in his daily pursuits which included hanging out in the theater, supervising the building of sets, and exploring under the house and shed in search of mice. It was a tough workday for him, so at night when he was in the house, he was too tired to be involved with the feuding females. All he wanted to do was sleep. Once or twice he put both of them in their place, in no uncertain terms, when the fighting intruded on him, but most of the time he simply ignored them.

Boundaries were established by Little Cat but Minnie rarely paid any attention to them. Her daily mission was to drive Little Cat crazy. Under the best of circumstances, any new addition to a household is a major adjustment for an animal already living there. Dogs are better at it than cats but eventually even the most hard-nosed cat will come around. I seriously wondered about Little Cat. She had been the queen of her realm for years and she wasn't about to make room on the throne for a loud, pushy, ghostly and obnoxious newcomer. I thought I was doomed to withstand a hundred-years war. In cat years, of course.

The turning point in their relationship came when I got a job at a theater in Virginia for the winter. When the season ended at the barn in Pennsylvania, I went back to New York for a week or so and then, with the car piled high with clothes and the animals, I headed south.

In those days, there were more venues for an actor, that is, for actors who hadn't made it in New York or Hollywood and I certainly fell

Speaking of Cats

Above: Little Cat and Minnie. Below: Little Cat and Nutsy.

Speaking of Cats

A chilly day in the Beach house in Virginia. Little Cat, Minnie, and Nutsy.

Nutsy after his accident.

Speaking of Cats

Bridie and Finn.

Bridie.

into that category. There were dinner theaters and stock and repertory companies all over the country. There still are some, but they are few and far between. Television killed off most of them. Why get dressed and go out to, perhaps, experience the excitement of a shared epiphany watching a play with live actors, when you can isolate yourself at home and watch reruns of mind-numbing shows while you eat everything in the refrigerator and scratch wherever you darn well please? It's pretty frightening.

Little Cat was an excellent traveler in the car. It took her only a few minutes to settle down and a find a good spot to sightsee and there was never a sound out of her. I wondered if that was because a car was her first secure environment after I found her on the road. Whatever the reason, she seemed to love it. Nutsy was something else. He hated riding in the car and in true Siamese fashion was very vocal about it. Even with the windows rolled up I'm sure people in passing cars could hear him complain. Anyone who has ever traveled with an unhappy Siamese will know there is no way to describe the unimaginable horror of it. And he had another problem that added to his displeasure. He suffered from motion sickness, so aside from the howling, I had to wait for the inevitable barfing. Not a pretty picture no matter how you look at it. But once he got sick, which meant pulling over and cleaning up, he settled down in some dark secret place between the luggage in abject humility and embarrassment and went to sleep for the rest of the trip.

Minnie had only been in a car a few times and the minute the door was shut she started acting as though aliens were abducting her. Between her and Nutsy, before he got sick and settled down, the decibel level of Siamese caterwauling was higher than any punk rock. Even Little Cat couldn't stand it and much to my surprise, went to Minnie, who was perched on the top of the back seat with her ears pinned down and her bugged-out eyes wide with panic and hysteria. I wasn't sure what Little Cat was going to do and I almost ran off the road trying to watch traffic and the two cats in the rearview mirror. After a moment, she sat close to Minnie, so close that they touched and I realized she was trying to console her. Clearly, this had to be a very conscious effort because prior to that day Little Cat wouldn't get within ten feet of Minnie if she could avoid it. There was no other explanation; it was a first and very impressive step toward calling a truce. "Adversity is not without comforts and hopes," said Francis Bacon. That certainly was the case for poor, crazed Minnie. Little Cat even gave her a quick lick or two, assuming Nutsy's parental responsibilities while he was indisposed. We still had an hour or two of intermittent screaming but Little Cat didn't give up on her nurturing. Finally, I think because Minnie was so hoarse she couldn't make another sound, there was blessed peace and quiet in the car. It was so quiet I could even hear the horns of the tractor-trailer trucks that were blasting at me to get out of their way.

Little Cat sat with Minnie for most of the trip and by the time we got to Norfolk they may not have been the best of friends but they had moved to a new plateau in their relationship. It was like a very shaky arrangement had been arrived at and both parties were doing their best to live up to it.

I rented a house on the beach in Sandbridge, Virginia, which was very reasonable because it was off-season and so few houses were occupied. I think the owner just wanted someone in the house when there were so few people in the community to help protect it from vandals. It was one of the best winters I can recall. And the animals liked it as much as I did. The ocean was a new world to them. There were sand dunes to climb and sand crabs to chase, gulls and pipers to watch and fishermen in the backyard casting their lines. There was never a dull moment. Little Cat and Nutsy were smart enough to stay away from the surf but Minnie was so curious and so fearless that more than once she was shocked when she found herself knee deep in salt water. Cats are so funny when they get unexpectedly wet. They do that wonderful "I hate water" dance where they walk away shaking one foot at a time as they go. And Minnie was even funnier because her hind legs were slightly bowed.

For most of us the beach is a jumble of summer images. Sunbathing and wet towels, gritty bathing suits, lotions, junk food, transistor radios and painful burns. But winter on the ocean isn't part of our lexicon and discovering how extraordinarily elemental it is, how fierce and unpredictable it can be, was one of the great joys of my life. And, as I said before, I know the animals enjoyed it. There's no mistaking how an animal feels, even a cat, who has

Speaking of Cats

a reputation for being secretive and stoic and removed. If you ever see a cat race around the garden, up a tree, down a tree, on the roof, off the roof, with the utter abandon that can only be motivated by the sheer joy of being alive, words like secretive and stoic go right out the window. All three cats, Nutsy, Little Cat and Minnie, would dash out of the house in the morning and run up to the top of the dune to see what the new day had to offer and what gifts the ocean had deposited during the night. There was always something interesting. I remember one morning finding a sea turtle about as big as a turkey platter. Why it was so far from the surf I didn't know but I'm sure we were much happier to see it than it was to see us. I turned it around so it would face the water and it awkwardly made its way back into the ocean. The cats were fascinated. There was no hiding their expectation and excitement because it was all too intoxicating. Secretive? I don't think so. Even in the snow and rain ...if it wasn't a downpour ...they forced themselves to check out the action on the beach.

For me, one of the most extraordinary and attractive qualities in animals is their highly developed sense of acceptance. What is, is ...

whatever the season, whatever the weather and whatever the time. And their reactions to their circumstances are always honest because they don't pretend. They don't know how. I'm sure that's a great part of their appeal to people. They don't lie, they're nonjudgmental, they lavish you with unconditional love and are happy to greet you every morning no matter what you look like. Who could ask for anything more? And they don't wear masks. Good, bad or indifferent, they are who they are. It's that simple. It seems to me there's a great deal to be learned from them.

In early spring I decided that it would be nice to have Minnie become a mother. I wasn't interested in going into the cat breeding business but several friends had said they wanted a kitten if and when Mrs. Fiske gave birth. The timing of her heat seemed perfect; she'd get pregnant and have the kittens soon after we got back to Pennsylvania for the summer season. I met someone who knew someone who had a very beautiful Siamese tomcat and I arranged a liaison. Whether or not he was the man of her dreams, I don't know, but he was a very handsome fellow and after only one date ... no flowers, no candy, no wine ... she was pregnant. It's an old, old story.

And so, as the theater season in Virginia was winding down, Minnie was blowing up. She took her pregnancy in stride, enjoying the newness of the early spring on the beach, still getting wet feet and chasing sand beetles. Life was good in Virginia ... so good, in fact, she decided her children should be born there and three days before we were to start out for Pennsylvania, she gave birth to four beautiful kittens. This was not according to the plans I had made but Minnie couldn't have cared less. It was time and she gave birth and that was all there was to it. She obviously enjoyed the birthing because, remarkably, she purred all through the delivery. The logistics of getting her, her kittens and the rest of the family back to Pennsylvania was my problem. She had her paws full with four babies that needed to be looked after.

Once again, with the car jammed full of luggage and animals, and this time, a nursery box, we hit the road. Little Cat settled into her usual place, right behind my neck on the back of the driver's seat. Nutsy started screaming in preparation for throwing up, which he did, inconveniently, while we were crossing the

Chesapeake Bay Bridge. Among all his other attributes, he was dependable. The one noticeable difference on that little journey was Minnie. She was blessedly quiet. She was too busy in her box, feeding and tending to the kittens. This is just a suggestion, but if you have to travel a long way with a female Siamese cat, you might think about getting her pregnant in time for her to deliver just before you leave. Looking after a litter of kittens is so time consuming and exhausting they don't have time to complain. It's better than valium.

Eight weeks later, when the kittens were old enough to leave home, their new families were delighted to get them. Both Minnie and I were delighted to see them go. As charming and endearing as they are, when they get old enough to start to explore the world on their own, four kittens can be exasperating. They seem to take turns sleeping so that at least one of them is awake to get into things. I grew up with twin sisters and my mom always said, as children, what one didn't think of to get into trouble, the other did. Well, Minnie had four babies and by the time the last of them went we were both worn to a frazzle. Nutsy and Little Cat weren't at all unhappy to see them leave.

Speaking of Cats

There was a collective sigh of relief when the last of them went out the door and we settled back into the complacency of our daily lives. None of us suffered the empty-nest syndrome.

I had decided that one litter was enough for Minnie. We had all shared the experience and even though it was enlightening, illuminating, uplifting, enriching and all those other adjectives that are usually connected with birth and kitten rearing, once was quite enough. But Mrs. Fiske had other ideas. When I took her to be spayed, not very long after the kittens had left, the vet told me she was pregnant again. How was that possible? She hadn't come into heat as far as I knew, was hardly out of the house, there was no tomcat around but Nutsy, and he had been neutered when he was a kitten.

"There's no way that cat can be pregnant," I said hysterically to the doctor. "No way at all. Do you hear me?" I all but grabbed him threateningly by the lapels of his white coat. "No way!" But she was. She had met a travelin' man and, well ... what can I say?

Before the discovery of Minnie's tryst and eventual pregnancy, I had made plans to go to Ireland, England and Scotland for three weeks soon after the summer theater closed. I was flying Icelandic, a very popular airline for actors and other bargain hunters. The reservations were already made and paid for and I had arranged for a friend in New York to feed the cats and clean their litter while I was away. Needless to say, when I made the arrangements, Minnie wasn't pregnant. I didn't know how my pet sitter would take to the added responsibility.

Minnie waited until she got back to New York to give birth so she was free to scream for the whole eight-hour journey. As usual, Nutsy carried on until he got sick and settled down but Minnie didn't quit. Even Little Cat gave up on her. Once they were back in the apartment, on home ground, everything was fine and a few days later Minnie had seven shiny, beautiful black babies. Again, she purred the whole time. The kittens were healthy and active but there was not a Siamese marking on any of them. There were no patches of orange, brown or white. They were pure witches-broom, Halloween-cat black.

I worried about Nutsy and Little Cat bothering the kittens while I was away, so another friend agreed to take them until I got back. The original cat sitter was happy to come in every day to feed Minnie and her babies, so I left for Europe with a light heart.

I remember the plane stopping at Keflavik airport in Iceland to refuel and getting out to stretch my legs. It was mid-October and I expected it to be much colder. Well, after all, it is Iceland. Actually, it wasn't much colder than New York. The passengers went into the restaurant to get a snack and just as we were settling at our tables, before I had time to put sugar in my coffee, we were asked to leave. Shocked, we dutifully filed out and stood with our noses pressed against the glass as a plane-load of Russian dignitaries and their aides marched in without so much as a glance our way. We were all incensed, European as well as American passengers, but it was during the Cold War and was a rude awakening, especially for Americans who had lived in a safe, myopic world that hadn't seen fighting on home territory since the Civil War.

My holiday was terrific and I must admit I hardly gave the cats a second thought. There was just too much to do and see. I had been to England and Scotland when I was stationed in Germany in the mid-fifties, so my love affair with both had started early on, but this was my first trip to Ireland. It was all I expected and more. The country is astonishingly beautiful, as comfortable as a rumpled eiderdown. The people were hospitable and generous and I was made to feel at home wherever I went ... with one notable exception. A truck backed into my rental car, denting the fender, in the midst of a livestock sale that seemed to be as crowded as Times Square. But when the red-headed, freckle-faced policeman, who looked to be in his very early twenties, arrived on his bicycle, nervously asking questions, there wasn't a single witness to the mishap. Nobody saw a thing and every friendly face turned stony-cold and looked at me accusingly. We no longer spoke the same language. Amazing!

All the cats were alive and well when I returned to New York, but the kittens were not what I had anticipated. They were wild things. The friend who looked after them only came in to feed and clean up, so most of the time their only interactions were with their mother. They were similar to feral cats, those mysterious flashes we see in barns, in the debris of cities and crossing dark streets in the early hours of the morning. Cats that only have themselves to depend on are so far removed from pampered and beloved pets, that they seem almost to be a different species. Minnie's kittens ran and hid whenever I came near them. In fact, they were so good at hiding that I hardly ever saw them. When it came time to find homes for them, I had an impossible time catching them and when I did they bit and scratched me. And why not? I was a threat to them.

I decided to keep one of the kittens, a sleek and beautiful little girl, that I named Rosemary. The spooky movie, *Rosemary's Baby*, was a big hit at the time and this kitten was certainly spooky. But, try as hard as I did, she never trusted me. She spent her life in the shadows, watching but never participating, and was happiest when she was in the country where she could disappear for days without being bothered by human contact. She maintained a good relationship with Minnie and the other cats and dogs, but people were another matter altogether. The odd thing was, when I did manage to catch her and pet her, she purred louder than any cat I've ever lived with. She wanted love so much but didn't know how to give or accept it. Isolation is deadly. As we all drift irretrievably, further and further into the impersonal realm of cyberspace, alone with only our computers to touch and no one to touch us in return, I wonder what we'll be like in a hundred years. I think of poor, solitary and forlorn Rosemary living in her lonely world and unable to come to the table to enjoy the feast which is our birthright to savor ... the shared experience of living.

> One of the most striking differences
> between a cat and a lie is that a cat
> has only nine lives.
> —Mark Twain

BRIDIE AND FINN

Minnie, being the youngest of the three cats, was the last to survive. Rosemary was younger, of course, but she came and went as she pleased. One day she went and never came back. She had already taken to being gone for weeks at a time so her disappearance wasn't noticed for quite a while. Finally, I realized she wasn't coming back, so I decided to always think of her in some splendid cat Arcadia and to this day I still do. Rosemary will live forever. I still expect to see her on the wall outside my kitchen window where she used to sit watching what was going on in the house, but not quite knowing how to be a part of it.

Minnie, on the other hand was always with me, always vocal and always lovingly amusing. Being the last of the triumvirate, she was immediately and sorely missed when, as quite an old lady, she died. In some way she was the end of an era and after she was gone, for whatever reason, I decided to live without cats.

I was already living in this house, which is next to the last on a quiet country road. Well, not too quiet. We may not have had to deal with street noises, but back then there were plenty of animals to make a racket. Dogs, chickens, ducks, geese, turkeys and guinea fowl, all of them greeting the day, chasing one another, or squabbling over a juicy bug. It was a loud and lovely cacophony and it eventually attracted the only cat that lived in the neighborhood.

I'm reminded of a very peculiar thing that happened when I had all the fowl running around the place. Bantam hens enjoy being mothers so much they'll sit on any eggs they can get when they're broody. They'll also sneak off and lay eggs in a nest you know nothing about and three weeks later suddenly appear with a dozen or so chicks trailing behind them. That's exactly what happened with one saucy little hen that was particularly addicted to motherhood. The father, or fathers, of this particular brood could have been any of the roosters strutting about the place so her chicks were all colors of the chicken rainbow. For a reason that will remain a mystery forever, four little cockerels, when they were old enough to be on their own, decided to strike out and move up the hill to my friends and neighbors, the Millards. Why just the four of them and not the rest of the brood, I'll never know. It was a risky business for the young chickens because Susanne and George lived with an Irish terrier named Magee who chased coyotes, bears, rabbits, squirrels and anything else that wandered onto her property. Yet she never bothered the roosters who eventually became known as the Gang of Four. The dynamics of animal relationships, in this case, why the chickens trusted Magee and why she ignored them, are fascinating. If only they could talk. The gang rarely came down to visit me, spending most of their time in the Millard's orchard eating insects and seeds and other chicken delectables. On occasion they appeared at their bird feeder. Every afternoon at teatime they paraded into the yard where they drank water out of Magee's bowl and then went on about their business. They slept in the trees, the four of them wing to wing, through rain and snow. As a result their feathers were beautiful, always as clean as if they had just taken a bath. They were inseparable and if one decided to visit my flock the four of them showed up. On a particularly glorious autumn day, when they were in my yard, the red member of the gang walked into the Hav-A-Heart trap I had set to catch whatever was raiding the hen house and eating eggs. I didn't check the trap until the following morning and there was the little rooster frantically trying to escape. Of course, I knew where he wanted to go, so I took him to the front of the house and released him. He started up the driveway squawking what I'm sure, translated from chicken was, "I'm free, I'm free, I'm free." The other three members of the gang heard him and came racing out of the orchard. I haven't seen anything as touching as that reunion since Scarlett managed to get back to Tara. They were all talking at once, dancing around each other as though they were welcoming a long lost brother home from the wars. It was remarkable to see their obvious affection for one another. Chickens! Go figure.

But, this is a book about cats.

I had seen the cat that was eventually attracted to the noise of the fowl, sitting on the porch of a house down the road. It looked to be about three months old, gray and fluffy and very much like the first cat I lived with, the mysterious Socrates. When I noticed the cat in my garden it was sitting on the split rail fence watching a hen scratch in the needles under an enormous old pine tree. There was nothing menacing about it, it was simply curious about the chicken activity. I would see the cat more and more until it started to spend a good deal of the day at my place. Not knowing how the neighbor felt about this I stopped to tell her the cat was at my house in the event that she was looking for it. She wasn't. As far as she was concerned the cat could stay away forever. It was running up the drapes and up her leg and she had the scratches to prove it. Teen- age cats will do that. It's probably the same hormonal insanity teenage kids suffer from. She said her children liked it but they were in school all day and didn't really miss it when it wasn't around.

Speaking of Cats

The cat's name was Michelle.

And so, I went home knowing full well my decision to live without cats was losing its resolve, and there was Michelle sitting on my patio as though she was waiting for a decision about her future. Once again it was fate so I invited her in and, without a second thought, she accepted my invitation. As she crossed the threshold I noticed that Michelle would have to be re-christened. She was a he.

This was a fearless young cat. His human counterpart would have been a boy who always had skinned knees and Band-Aids, a dirty face, a baseball cap askew on his shaggy head and used "bad" words. He had a healthy appetite and an insatiable curiosity, which got him into a lot of trouble. He was like a street kid or maybe Huck Finn. Yes, Huck Finn was more like it. So, Michelle became Finn and the name suited him very well.

A few months after Finn settled in and became a member of the household, I was having dinner with the Millards, where the Gang of Four lived, sitting on the point behind their house where they have a spectacular view of the mountains and canyons. This is a view most people can only dream about and, strangely, always reminds me of the timeless, evocative feeling I have when I'm on the beach looking at the ocean. Even on the hottest night there is always a cool breeze off the mountains, so that night we were lingering over our coffee and settling all the problems of the world, when I felt something brush up against my ankle. Startled, I looked down and there was a very little, tiger-striped brown and gray kitten. It couldn't have been more than eight or nine weeks old. I picked it up, got a good look at it and noticed it was a little girl. The only place it could possibly have come from was a house on the rim of the canyon, above the Millards, which was only accessible from another road. I knew the owners of the house and also that they had cats but never let them indoors. Very dangerous in coyote country. I didn't need another cat and I didn't want another cat but she was very little, seemingly lost and she reminded me of Little Cat. However, she lived with someone else and even though her chances of surviving the predators were slim I couldn't just take her home. I looked at her sitting comfortably on my lap and decided that if, by chance, she followed me home ... well, it would be fate telling me she should live with me. So, I took a teaspoon of milk from the milk pitcher and fed her just in case she hadn't already attached herself to me and petted her until I was ready to go home. Could I simply carry her off? No, no, that would be dishonest, stealing, catnapping ... so I put her on the ground and started for home. It's about a city block from the Millard's house to mine and I knew her little legs would never be able to keep up with me if I walked a normal pace so I took baby steps all the way home. It must have taken me twenty-five minutes but she was with me every inch of the way. When I got to my driveway, I bid goodnight to Susanne and George and figuring it was close enough I picked her up and went inside. Well, what was I to do? She was homeless, little and hungry. I had to do the decent thing and take her in.

Honesty is in the eye of the beholder.

She was a wonderful kitten, as different from the pugnacious Finn as a cat could be. She was playful, gentle and very loving and the more I got to know her the more she reminded me of Little Cat. I was tempted to call her Little Cat, Too, but decided she needed a name all her own so I started the process of trying to come up with exactly the one that suited her. For weeks she was just Cat, like the big orange cat in *Breakfast at Tiffany's*. And then, one day when I was talking on the phone to a dear friend, Madeleine, a tiny, elderly French woman who was feisty, outspoken and filled with so much curiosity she was always a joy to spend time with, I found the perfect name. She was talking about her relatives in England, members of her late husband's family, whom she wasn't particularly fond of, and she mentioned the one she did like. Her name was Bride. Madeleine repeated the name several times in the course of the conversation, Bride this, Bride that, Bride, Bride, Bride and all the while I was sitting there with the nameless kitten in my lap. It was the perfect name for her. Bridie.

And, so, the new cats living with me, after I had decided not to share my life with cats, were Bridie and Finn.

I was writing for television at the time, a precarious profession at best, but I was past fifty and I knew my days were numbered. I was a dinosaur well on my way to extinction, about to be fossilized in whatever strata old writers get mired in. There are three things one cannot be, under any circumstances, in Hollywood. Sick, broke or old. No leper was ever shunned as much. No one wants to be reminded that sooner or later one condition or another is going to get every last one of them. Mortality, on any level, is a whispered dirty word.

I wasn't ready to spend my days going to the senior center for tea and tai chi or riding around on a golf cart deluding myself into thinking I was getting exercise, so while I was still working at the studios I started a novel I had had bouncing around in my head for several years. It was to be an homage to my childhood, not exactly an original idea for a first novel but, more than anything, I wanted to see if I could do it. *Livery Street* was the title I decided on; a fictitious name for the street where I grew up in an Irish-American neighborhood in an Ivy League town. The narrator of the story was Tim Gallagher and his nemesis was Maureen O'Connor. It was not autobiographical but I wanted to capture the ambiance of what it was like before we, as individuals and collectively as a nation, lost our innocence. And, I wanted to write about the magic of growing up in a real neighborhood where we all shared in the very good and very bad times.

Since I was still working I had scant time

to devote to the book, weekends and evenings if I wasn't too exhausted, and more than once I decided to abandon it. But friends read bits of it and were very encouraging so I stuck with it. After awhile, the title, *Livery Street*, wasn't working for me. I couldn't picture it on a dust jacket, so I changed it several times and finally decided on *Maureen and Tim*. It seemed to personalize the story and I hoped it would for the readers ... if there would ever be any. But, as a friend pointed out, in an age when personal relationships were being examined inside out, up, down and sideways, *Maureen and Tim* sounded a bit like a married couple on very shaky grounds trying to work out their problems. With that thought planted firmly in my insecure mind I started looking for another title and wondered how long it took Jane Austen to come up with *Pride and Prejudice*. And then, one winter day when the sun was streaming in the windows of my office, warming the cats who were curled up on the deep sills, when I had long since put the search for a new title to rest, it hit me. Instead of calling the girl Maureen O'Connor, I'd change her name to Bridie O'Connor, and Tim Gallagher would become Tim Finnegan. And, that was the new title of my book. *Bridie and Finn*. I liked it. Fate.

Perhaps Jane Austen lived with two cats named Pride and Prejudice. Or maybe Sense and Sensibility. For certain there had to be a cat she called Emma!

I have to include this little anecdote because it always comes to mind when I think of Bridie and Finn ... the cats and the book. My agent called me one day, before the novel was published, to tell me Bridie and Finn had been chosen for the WHSmith Fresh Talent Promotion in Britain. Naturally I was excited and asked, "What does that mean?"

"I have no idea," she answered, "but it's a good thing."

WHSmith, the largest retail booksellers in the United Kingdom, choose six first novels every year to promote in their shops. It's a very good thing, indeed. I was one of two Americans picked that year and to me it was like winning the Nobel Prize. For promotional purposes they wanted a photo of the six authors together so I was flown to England where I was put up at a luxury hotel and wined and dined for a week ... just to have my picture taken! It was enough to make a grown man believe in Santa Claus.

My seatmate on the plane going over was

a young man of twenty on his way home to London after spending three glorious months with his beloved and much admired older brother who lived in New Zealand. He was going back to college and although he was anxious to see his parents he wasn't happy about leaving his brother and he was drowning his sorrow in too many beers. There was no end to the stories he had to tell about his stay in New Zealand.

"I did a little sheep shearing. It's a lot harder than it looks. And I went bungie jumping. Scared the stuff out of me but it was brilliant. I did it every chance I got. And I saw Prince Charles. I live right in London, I have all my life, and I had to go all the way to New Zealand to see Prince Charles! Fancy that!"

We talked on and on, even after the flight attendant decided that he had had quite enough beer and I suggested he switch to coffee. I didn't want to see him stagger into the arms of his anxiously waiting parents at Heathrow. Finally, after exhausting his stories about the adventures of he and his brother, he asked why I was going to London.

"On business or a holiday?"

"Actually, both. I'm going over to have my

picture taken." He was very impressed when I told him what the picture was for.

"You're a writer? That's brilliant. Bloody brilliant. I want to be a writer." He shook his head in wonder. "What a lucky bloke I am. And what a brilliant holiday this has been. First Prince Charles and now, you!"

The Cat in gloves catches no mice.
  —Benjamin Franklin

NUTSY

Before I go any further, I have to say Nutsy was the greatest cat that ever lived. And that's an understatement. It's the truth. Now, I know that anyone who has ever had a cat they dearly loved will argue that theirs was the greatest and I wouldn't expect anything less, but Nutsy was a cat amongst cats.

He was born in Manhattan in the heart of Greenwich Village. I had already met Fannie Flagg's Siamese cat, Tallulah, and was taken with her good looks and wackiness. The cat's, I mean. Well, Fannie's, too, for that matter. So, I started looking at ads in the paper until I found the one that eventually led me to Nutsy. I called and the woman gave me her address and we arranged for a time to meet. She lived on the fourth floor of a walk-up in one of those tiny apartments that made it possible for artists to survive in New York. The woman was typical of the people who were "new age" back in the days when "new age" was already old hat in the Village. She was probably in her mid-forties, all smiles and bright eyes and gray hair hanging loose down to her waist. She wore no makeup but every digit, including her neck, wrists and earlobes were adorned with some kind of jingly, jangly jewelry. She couldn't make a move without sounding like wind chimes. Andean flute music was playing softly and mystically in the background and the first thing I thought of was a fortune-teller, what with the beaded curtains, very peculiar dolls and astrological signs everywhere. The walls were painted bright red and the woodwork was black and there were candles burning on almost every flat surface. Nothing masks dirt and imperfection as well as candlelight. When she opened the door the mother cat gave me a quick ankle polish before trying to run into the hall, but the woman grabbed her just in time and greeted me.

"I only have one kitten left. A little boy. But he's very sweet. He's a Capricorn." Her eyes widened. "You know what they're like."

"Oh, yes." I didn't, but I took her word for it and assumed it meant there was something special about him. Actually, I was glad there was only one left because I couldn't imagine having to choose one over the other.

"You have a warm aura," she said, smiling even more broadly than she had before. "Very warm. The two of you will be good for each other." It was probably the same sales pitch she used with every prospective customer but it's always flattering to hear nice things about one's aura.

"Thank you."

"Where did he get to? He's my favorite. The only male in the litter. Seems everyone wanted a little girl," she said looking under the furniture. "Come on, puss. Come on out. You have company." She stopped in the midst of her search. "Would you like some chamomile tea? It won't take a minute."

"No thank you. I'm kind of in a hurry."

"There you are." She reached under the sofa and pulled out a very leggy kitten of about nine- or ten-weeks old with that wonderful pop-eyed look animals get when they're being introduced to something or someone new. He was irresistible. Seal point Siamese kittens, before they have their full coloring, appear to have stuck their nose and feet into dark chocolate and there's automatically something amusing and silly about them. And, I might add, very endearing.

"I've been calling him Moon Baby. Moon Baby," she repeated as though she expected something miraculous to happen when she said the words. She noticed my lack of enthusiasm. "I suppose you'll want to change that."

"Probably." By then I was holding the kitten and he certainly was no Moon Baby. "Of course, I'll take him. You said twenty-five dollars?"

"Yes. I wish I could just give them to loving parents but … "

"Oh, no. Twenty five dollars is fine." I gave her the money and headed for the door.

"Just a minute." She took the kitten from me and cuddled and kissed him. "Bye, baby. You be a good boy." She handed him back to me. "Would you like me to read the Tarot cards? No charge, of course."

"Gee, maybe some other time. I have your number."

"I'm very good. Just call whenever you like. And tuck him into your coat. It's cold out."

"Thank you," I said pulling the lapel over him as I started down the stairs.

"He's litter trained," were her last words to me.

He had never been out of the apartment and the noises of a crowded New York street were making him very nervous. I could feel him shaking against my chest and when he'd peek out of my coat to look at this new and terrifying world, his eyes were wild and his ears were plastered flat against his head. I said every reassuring thing I could think of, even resorting to calling him Moon Baby. I worried about attracting attention, walking along the street talking into my coat, but it was, after all, Greenwich Village so no one gave me a second glance.

By the time we got to the subway station he seemed to have settled a bit but wouldn't even look out at what was going on.

Imagine how frightened a little animal is

when it's taken away from its mother, the only environment it has ever known, and thrust into an alien world. It's a wonder they survive. A part of me wanted to take him back to the fortune-teller, but I knew that sooner or later he'd have to go through this again and since we were already at the train stop, I put the token in the slot and went to the uptown platform. What I hadn't reckoned with was the effect the noise of the train would have on him when it roared into the station. It was disastrous. By the time the train stopped my shirt was shredded and my chest was a bloody mess. Kittens have sharp, needle-like claws and for a moment I didn't know whether to get on the train or go to an emergency room. I decided riding fifty blocks with an hysterical kitten in my coat, as the train thundered and rumbled and groaned while it swayed from side to side, wasn't a very good idea. So I went upstairs, left the station, did the unthinkable and hailed a taxi.

By the time the cab arrived at my apartment building I had decided to call the kitten Noah. It was a strong name, pleasant to the ear and a considerable improvement over Moon Baby. He settled into life on First Avenue very quickly and immediately bonded with the boxer, Phoebe, I was already living with. In no time they developed a symbiotic relationship. If they were alone in the apartment and there was anything on the kitchen table, the counter, on top of the refrigerator, anything at all, Noah knocked it to the floor and Phoebe ate it. She did not have a discriminating palate. Whole loaves of bread, wrapper and all, disappeared down her gullet. She once ate the stuffing out of the front seat of Nancy's cream and green Chevy when she was left waiting for no more than fifteen minutes. It was a lovely cool day, the windows were open so there was a breeze, but she got bored and tore into the seat like a demolition crew. Noah watched the whole thing and, apparently, never uttered a sound.

Actually, he was Noah for a very short time. His lack of agility was soon very apparent. He'd do things like jump up on the coffee table and accidentally knock everything off or he'd decide to explore the bookcase and clear a shelf or two. No doubt about it, he was a klutz. And in true kitten fashion, he did everything without thinking. Once, when I was relaxing in the bathtub he exuberantly raced into the room and without thinking, jumped in with me. Horrified, he put his front paws on the rim of the tub and, dripping wet, looked at me as though I was out of my mind for sitting in water. He jumped out, shook himself and walked away doing the famous wet-foot jig. He was a nutsy kind of cat and so, Noah became Nutsy or occasionally, Mr. Nutser.

But what he lacked in agility and common sense, he more than made up for in obvious charm. He was the gentlest, most loving cat I have ever known, as well as the most intuitive. He was remarkably fine-tuned to moods, somehow knowing exactly when he was needed merely to comfort. When I had serious problems with my lungs and was bedridden for a considerable time, he stayed with me all the time. With the exception of using the litter and eating, and I know he missed meals, he was always there. The same was true of any other illness I suffered ... Nutsy seemed to know and he was there. Had he been able to make a cup of tea, he would have been the best nurse since Florence Nightingale.

I've already mentioned what a good parent he was to Little Cat, letting her "nurse" on little tufts of his hair for hours when she was a sickly little kitten. Without a doubt he was responsible for her survival. To further illustrate the remarkable, compassionate, healing side of this wonderful cat, Nancy, who had had her car seat eaten by Phoebe and was abandoned on stage when Little Cat decided to quit show biz, swears she couldn't have survived a vicious toothache if it hadn't been for Nutsy. We were in the beach house in Virginia, far from any dentist, when late at night she developed a horrendous pain in her jaw. She took as much aspirin as she thought safe but it didn't ease the pain and by the time she finally went to bed she was in tears. Nutsy curled up next to her face, something he never did, with his warm body touching the cheek on the side of the bad tooth. He stayed there all night. The pain abated and Nancy finally went to sleep. As unbelievable as that may seem, to this day she swears she never would have gotten through the ordeal without him. Saints have been canonized for lesser miracles.

Nutsy was a perfect example of that classic character in literature, the lovable, bumbling gentleman. Guileless and sympathetic, devoted, warm and affectionate, Charles Dickens couldn't have fashioned a more beloved individual. I always had the feeling that he was a playful little man in a cat suit who only reverted to his feline ancestry when he had to ride in a car.

I, of course, thought he was quite a handsome fellow, but Siamese cat snobs referred to him, rudely, as an "apple head." That meant he didn't have the long pointy face of the classic Siamese and would never make it as a show cat, but one member of the family in show business was quite enough.

His klutziness and curiosity did, on occasion, get him into some pretty precarious situations. The owner of the house on the beach, whom I never met, had left a small bucket with motor oil on the screened-in porch. What with settling into the new theater, rehearsals, etc., I didn't get around to disposing of it for a few weeks. In fact, I forgot about it. Somehow, Nutsy knocked it over and when I came home from work, there he was covered with oil from his waist down. He was mortified. I quickly got the dreaded cat shampoo and gave him a bath, in spite of his ear-shattering protestations. After a few washes the oil didn't seem to be coming off. So, carrying him dripping wet, and I was as wet as he was by that time, I got my own shampoo and I gave him a few more baths. It seemed to work and when I thought he was as oil free as he was going to get, I dried him and let him go off to a dark corner where he could clean himself and soothe his wounded ego. Tub baths are infuriating and very embarrassing to most cats. They are downright unnatural and to be avoided at all costs.

About a week went by and Mr. Nutser didn't seem to be the worst for wear because of his trauma but I started to notice an unusual amount of cat hair on the rug. I knew it wasn't from Little Cat or Minnie. Nutsy was losing his hair at an alarming rate. I picked him up and could see the hair where he had been saturated with the oil, was very thin. And when I pulled lightly on it it came out in clumps. Having survived the shame of spilling the oil in the first place and the indignity of too many baths, he now had to contend with being denuded. And that's exactly what happened. I suppose the combination of the oil and shampoo caused him to lose all his hair and he spent the next several months naked from the waist down. He was, needless to say, a very strange looking cat. But, being the stalwart gentleman that he always was, he suffered the mortification with his usual nobility until, once again, he had his beautiful coat. Maybe he was just an "apple head" but royal blood ran through his veins.

Like most cats he enjoyed being outside, so

summers in Pennsylvania at the theater were probably his favorite time of the year. But cats, like dogs, want to be with their people. That's the most important thing in their lives, although they adjust very well to whatever environment provided as long as it's not cruel or harsh. I've never understood people who get dogs and chain them in their backyard for the entire life of the animal. That kind of treatment should be a felony as far as I'm concerned.

Nutsy and the other cats seemed quite content living in the apartments in New York. For one thing, they got more than their share of attention when the quarters were a bit more confining. And dozing on sunny windowsills on cold and blowing winter days was absolutely intoxicating for them. The three cats would pile almost on top of one another, and if it wasn't for their color you couldn't tell where one started and the other ended. Forget catnip, the sun was their drug of choice.

But, as I've said, summer offered the great outdoors and by comparison, the apartments paled. There was fresh grass to be gnawed at, something cats dearly love to do, and all kinds of creatures to chase. It truly was one long holiday, like kids being out of school until after Labor Day. But these cats, perhaps because they were used to sleeping inside, never took to staying out at night. They seemed to exhaust themselves during the day and were more than ready for a good night's sleep when the sun went down.

One night, just before the summer season started, Nutsy didn't come home. I went through the empty theater calling him but he didn't respond. He didn't seem to be anywhere in the immediate area. I went to bed more annoyed than worried, thinking that he was having a good time and not paying any attention to me. Early the next morning I went out and called, but still there was no Nutsy. Once it was daylight I walked behind the little house we were staying in and I saw what I thought was a rag, until I got closer and saw that it was him. He was covered with blood and his face was smashed in. I thought he was dead, but when I knelt over him I saw he was breathing. He was unconscious. I picked him up and with tears in my eyes headed for the car. It made me sick to think that he might have been lying there, suffering, all night not twenty feet away. I raced to the vet, which was about fifteen miles away, wondering what could have happened to him. There was no road in the back of the house. Surely, if he had been hit by a car and crawled to the house he would have gone to the front door. There was no door of any kind in the back so why would he have dragged himself there? All the way to the doctor's office I kept talking to him, hoping that in some way in his unconscious state he knew I was there and that everything was going to be all right. I didn't believe it for a moment and was certain he'd be dead before we arrived. The people in the vet's office, even the doctor himself, who were used to seeing that sort of carnage, were appalled at his condition. We took him into the examining room, laid him on the table and the doctor started checking him over.

"His jaw's broken in several places and he's bleeding from one ear. I can't tell if there's any damage to his eyes."

I started to explain that there was no road behind the house and wondered why he'd go back there instead of the front door. Half way through the explanation the doctor stopped me.

"This cat wasn't hit by a car."

"He wasn't? Then what could have happened?"

"Somebody hit him with some kind of club. I know it. If a tire went over his head hard enough to break his jaw, it would have killed him. This cat was hit with something. I'm sure it was deliberate."

I've never felt that kind of rage in my life. I didn't know whether to cry or scream or throw up. All three wouldn't have been enough.

"Is he going to be OK?"

"I can't answer that. We'll have to check him over carefully, X-rays, blood work … I don't know about any internal injuries. But I have to tell you, I don't think he stands much of a chance. I'm amazed he's still alive."

I could hardly get the words out. "Do anything … everything you can. Please."

"I will. Go home. I'll call you later today when I know what's going on."

"I can stay…"

"You can't do anything. Just go home. If I think he has to be put to sleep I'll call you before I do it."

On the way back to the house I kept asking myself, Who would do such a horrible thing? I knew how cruel people can be but this was an idyllic country town, a sleepy little hamlet in Western Pennsylvania, where everyone had some kind of animal and they all knew who each cat and dog belonged to. And there weren't that many houses close by. The big house, where the owner of the little place we stayed in lived, was in front about a hundred yards away but those people would never harm an animal. They were very humane and compassionate. Who would purposely club a cat?

I must admit, from that day on, for the next eight or so years that we had the theater, I never felt the same about the town. I didn't condemn everyone but there was some small part of me that was always aware that someone living there was evil.

That afternoon the doctor called and although the news wasn't as good as I had hoped for, it wasn't as bad as it could have been.

"His jaw was broken in four places and I wired it together. I'm still not sure about damage to his head but right now there doesn't seem to be any indication of severe trauma. His eyes may be all right. It's just too soon to tell. By no means is he out of the woods but … well, we'll have to wait and see. But, I don't think we have to put him down. Not yet."

"Thank you. Thank you so much. Shall I call later?"

"Tonight. Call about six o'clock. Maybe we'll know more."

At six he was a little better. Miraculously, it looked as though he was going to pull through, but he'd have to stay in the hospital for awhile. I didn't care as long as he was going to get better. And I was grateful to the doctor who was so concerned about him.

It was a week before he could come home. He had lost so much weight he looked like a skeleton. There were bits of wire sticking out of his mouth where the breaks were repaired. He was still a very sick cat and the doctor wasn't making any promises, but he thought if he was going to recover it would be better for him at home. When I could see the whites of Nutsy's eyes they were all blood shot. I was almost afraid to touch him because he was so fragile. He could barely walk and I had to give him medication and some liquid food with an eyedropper. The food didn't interest him at all. He seemed to want to sleep all the time. It was all he could do. I tried to keep the other cats and the dog away from him but it was impossible. The amazing thing was when they went to him to sniff those horrible medicinal smells, trying to figure out just what was going on, he purred. He was practically on his deathbed and yet, he purred.

The liquid food the doctor gave me wasn't doing it for Nutsy. He didn't even try to swallow and it just ran out the side of his mouth. So, I cooked some kidney, a favorite of his, mashed it and added it to the broth and he seemed to take a little of that. His poor thin body had been living off itself and he was wasted away to nothing. All he needed to get everything functioning again, was some food and the mashed kidney did it. It tasted so good to him that after awhile he tried to open his wired together jaw to eat. How incredible the body is! Cats, dogs, people, birds—it doesn't matter, once there is fuel it starts to do what it's supposed to do. Support life. Someone once said, "The body wants to live. That's its job." Nutsy was proof of that. Once he was getting nourishment he started to gain weight and his strength began to return. When the wires were removed and he could eat on his own, he was well on his way to recovery.

Speaking of Cats

There's a colloquialism I've only heard used in Western Pennsylvania, like those wonderful Yiddish words that describe something better than any normal verbiage ever could. The word is sywickered. The spelling is a shot in the dark because I couldn't find it anywhere. It describes something that is sort of crooked and a little skewed and slightly distorted. That was Nutsy's face for the rest of his life. Permanently sywickered. But among all the other attributes that make animals so special is their complete lack of vanity. Peacocks don't spread their magnificent tail feathers to show off just to impress anyone passing by. What they're doing is trying to catch the eye of a peahen. Those of us who by happenstance see them in all their splendor are just lucky casual observers. The peacock couldn't care less that we're there.

Nutsy had no idea that one side of his face looked like he was perpetually smiling and the other looked as though he was in a snit. The whole dreadful experience didn't change him a bit. He still greeted strangers as friends, as trusting as ever. But the sywickered face, if anything, made him more Dickensian. Dickens probably would have used the name. Lovable

Noah Sywicker, Esq., who wore a beige and brown swallowtail coat, a top hat and an embroidered waistcoat that might always have a crumb or two on it. Genial and gentle, slightly discombobulated Mr. Sywicker, who loved everyone and was curious about almost everything but who never left London because he got deathly sick when he had to ride the stagecoach.

AFTERTHOUGHT

Purposely, I have avoided dealing with the deaths of any of the cats I have written about. Once was more than enough and I choose to remember them sitting on sunny windowsills or curled up in the corner of a comfy chair. They all live on in my memory and always will. Denial can be a great comfort.

The model for the cover is the cat I currently live with. He is aptly named Bomber. Nothing more need be said. The ferocious look on his face, as though he's about to pounce on some unsuspecting creature, is a close-up of him, sitting on the table on the patio, watching me eat cornflakes and hoping I'll save him some. So much for feline inscrutability.

There's a hummingbird feeder on the patio and mornings when I eat breakfast there I take my camera with the zoom lens, determined to get one of those spectacular photos of a hummer in flight sipping nectar. So far all I've done is increase the value of Kodak stock and collect hundreds of pictures of blurs with long beaks. I'm discovering that, contrary to what the nuns told me all through grammar school, dogged persistence doesn't always pay off. Well, one morning when I was on yet another futile photo safari I scanned my immediate horizon and there was Bomber staring into my camera obviously waiting for his share of the cornflakes.

He came into my life when someone found him wandering around the parking lot of a busy restaurant. The man who found him had to go into Los Angeles, about eighty miles away, and wouldn't be back until late that night so, knowing I was a soft touch as far as animals are concerned, he asked if I'd look after the kitten until the next morning. I was more than happy to oblige. (As I write this it's five o'clock in the morning and Bomber, who has been asleep in some private little nook of his, has just jumped into my lap. I assume for editorial purposes.) The next day when the fellow came to pick up the cat it was too late ... he was mine. There is such a thing as love at first sight. Actually, the man, although humane, wasn't really that interested in having a cat live in his house so I was doing all three of us a favor. The moral of the story is, however, I'm not to be trusted.

Having read through these pages I'm aware of a pervasive feeling of guilt. While I was sharing my life with the cats I write about, there were always dogs and I've hardly mentioned them. I feel as if I'm not honoring their memory. To assuage my guilt I can only say that their personalities were equally as strong. Their love, devotion and charm were as captivating as the cats and had I gone into detail about them I would have confused myself and had to face up to my considerable limitations. And so, I apologize to Tucker, Jesse, Duke, Benjie, Delores, Biff, Amy, Gilly, Bing, et al. You're next.

Finally, I am struck by how puny words are. I haven't come anywhere near relating how privileged I was to share my life with these animals. Without them ... well, I know the sun would have risen and set everyday, the seasons would have dependably come around every year and spring would have happened just for me. And I know I would always have wondered at the magic of it all. But without the animals my life might have been like a black and white movie rather than in the glorious Technicolor that it should always be.

About the Author

Harry Cauley, a native of Princeton, New Jersey, is an award-winning screenwriter, playwright, and novelist. He has also appeared in more than fifty plays, and has written and produced dozens of shows for television. His novels include Bridie & Finn, The Botticelli Angel, and Millersburg. Mr. Cauley lives in California.